Bio Invaders

Invasiv   Marketplace ™

**SAVING NATURE
A to Z
By
Brett Scott**

www.bioinvaders.com

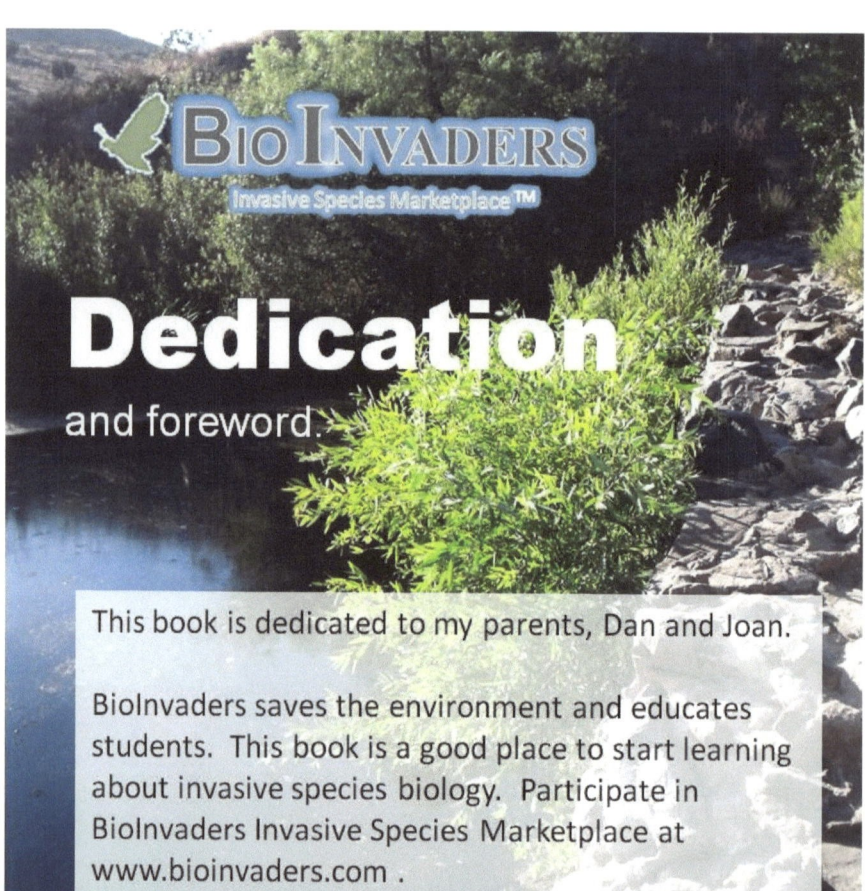

Dedication

and foreword.

This book is dedicated to my parents, Dan and Joan.

BioInvaders saves the environment and educates students. This book is a good place to start learning about invasive species biology. Participate in BioInvaders Invasive Species Marketplace at www.bioinvaders.com .

BioInvaders
Invasive Species Marketplace ™

G is for grow.

Growing plants, like this pineapple, is how we sustain life on Earth as they are an integral part of the food chain. Invasive pest species can harm agriculture which damages the economy and puts us all at risk.

H is for home.

Home Is where we live, work, and play.
Everyday starts and ends in the home.
Imagine if your home was overgrown by
Invasive kudzu plants like this one.

I is for invasive species.

Invasive species are non-native introduced species that cause harm in the ecosystem. There are many invasive species all over the globe, and there is something you can do about it. By reading this book you are helping the environment.

J is for job.

Jobs support families, and BioInvaders creates
as many jobs as possible. By buying this book
you helped create jobs. Invasive species harm
the economy, but we are here to help.

K is for kudzu stink bug.

Kudzu stink bugs ate my friend Virginia's Lady Banks Roses. We had to remove them from her garden and turn them into science experiments.

L is for ladybug.

Ladybugs are another example of nature's wonder. One of the most charismatic species, ladybugs can be found everywhere. If you find one, take a picture and submit it on www.lostladybug.org.

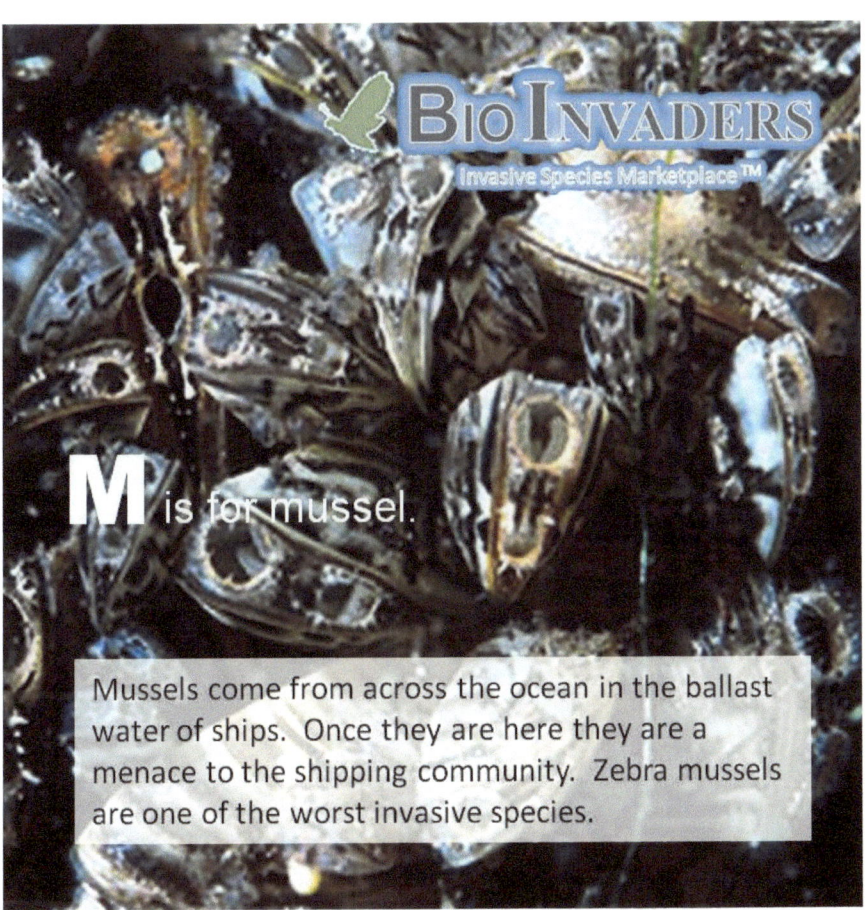

M is for mussel.

Mussels come from across the ocean in the ballast water of ships. Once they are here they are a menace to the shipping community. Zebra mussels are one of the worst invasive species.

N is for nutria.

Nutrias are a mammalian invasive species. We have them in Texas where this book was written. Usually, they have fleas all over them and can grow to be huge.

O is for origin.

Origins of species are of particular concern when it comes to conserving biodiversity. These tortoises are from the Galapagos Islands: a sensitive habitat that would be destroyed by invasive species.

P is for pest.

Pests are a danger to homes, gardens, and farms where they can do the most damage. Many invasive species are pests, like this snail, and scientists spend much of their time dealing with these.

Bio Invaders

Invasive Species Marketplace ™

BioInvaders Exploration of the Mayan Cichlid
Instructions: Identify the Parts of the Fish

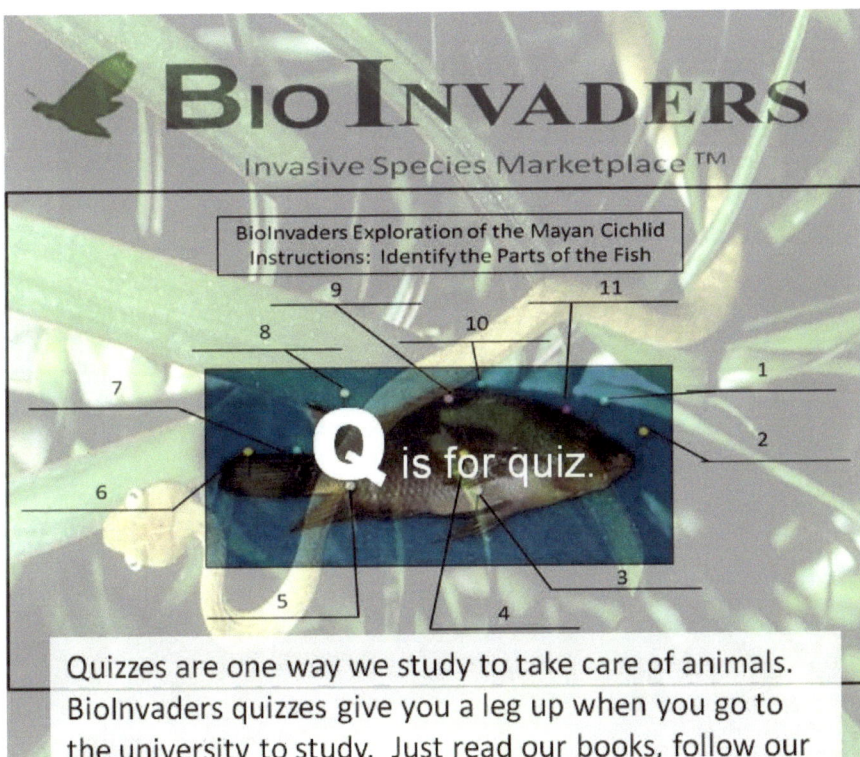

Quizzes are one way we study to take care of animals. BioInvaders quizzes give you a leg up when you go to the university to study. Just read our books, follow our lesson plans and study guides, and take our tests and you'll be prepared.

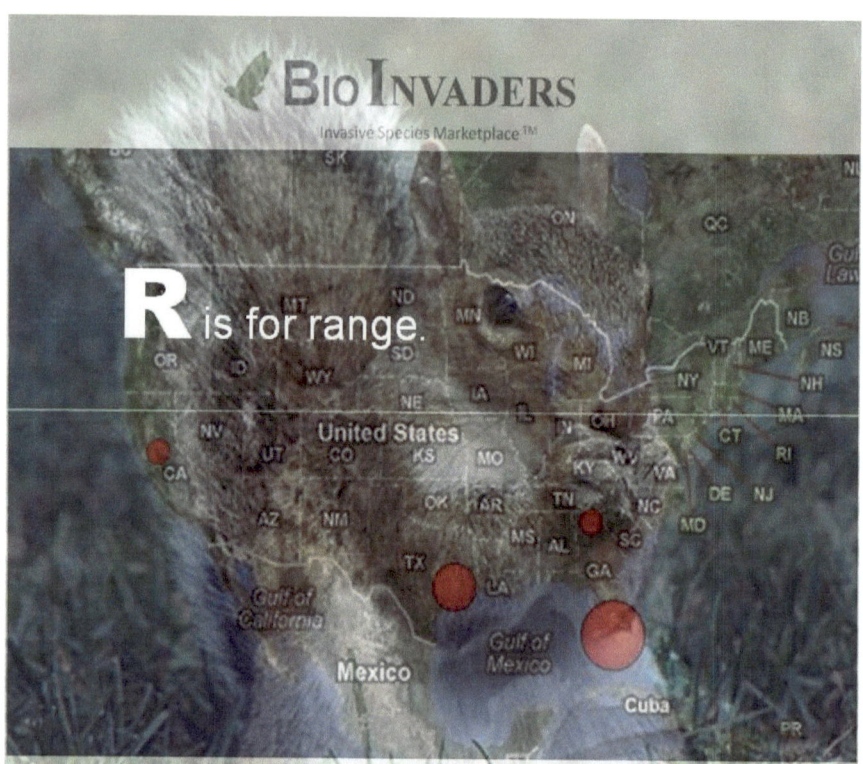

BioInvaders

Invasive Species Marketplace™

R is for range.

Areas where BioInvaders Incorporated is restoring ecosystems by collecting, sacrificing, and preserving invasive species specimens for use in the classroom or laboratory.

S is for sea.

Sea life is harmed by marine bioinvaders that do some of the most damage of any species. They can harm native coral species and native fish populations. Right now lionfish are one of the worst problem species.

T is for turtle.

Turtles are another special species. As they are one of our most precious creatures, we take care of them. Invasive species pose a risk to turtles by various mechanisms.

Bio Invaders
Invasive Species Marketplace™

U is for uberspecies.

Uberspecies are those extra special species that we all care about the most, like this monarch butterfly. Certain uberspecies are at risk from invasive species, and those are the ones we protect the most.

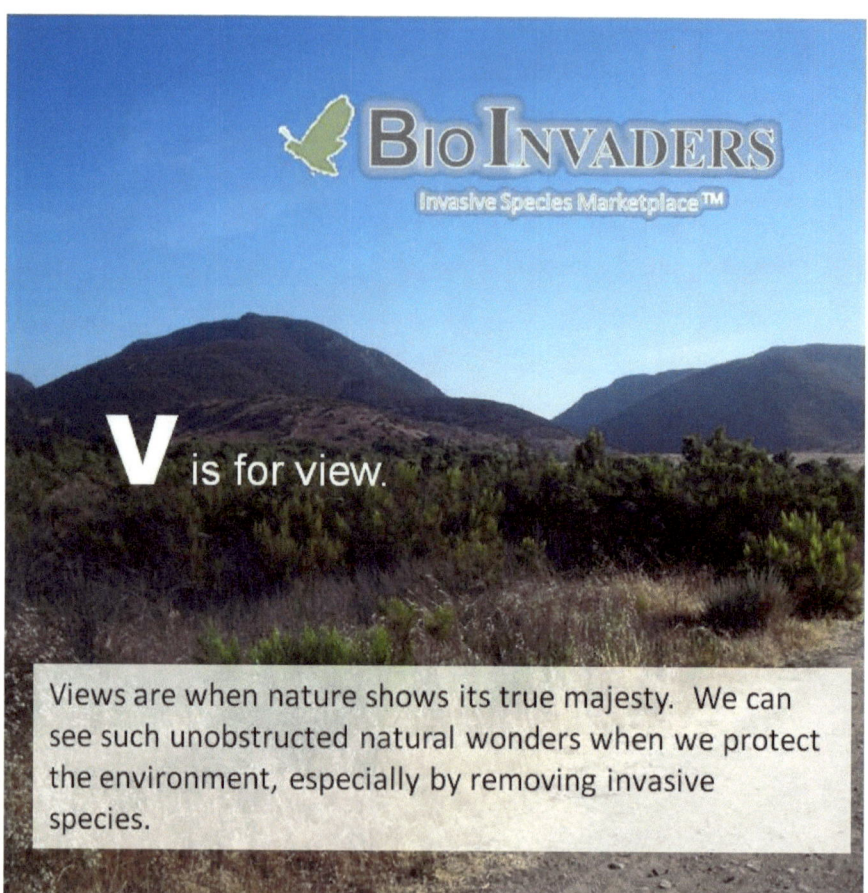

V is for view.

Views are when nature shows its true majesty. We can see such unobstructed natural wonders when we protect the environment, especially by removing invasive species.

W is for water.

Water gives life to the Earth. Various chemical characteristics of water provide a miraculous substance from which come all living things. Keep the water clean and don't put invasive species in there!

X is for xenopus.

Xenopus are a type of frog, and sometimes they can be invasive, such as this Xenopus laevis here. BioInvaders targets these organisms for removal from the environment and use in the classroom or laboratory.

Y is for young.

Young animals often require the care of their parents, like this baby chimpanzee. We remove invasive species from the environment to protect all the native young animals.

Z is for zoo.

Zoos are home to some of our most spectacular creatures like this white tiger. Zoos are one of the safe places where humans can observe captive wild animals. Visit the zoo and bring this book with you and see how many animals you can spot.

Brett Scott, MBA, President of BioInvaders
Studied Biology at Cornell University and
Business Administration at University of California,
San Diego.

www.ingramcontent.com/pod-product-compliance
Lightning Source LLC
Chambersburg PA
CBHW041119180526
45172CB00001B/330